Plants and Plant Growth

W9-AWG-000

Printed in Mexico

ISBN 978-0-15-362074-4
ISBN 0-15-362074-9

2 3 4 5 6 7 8 9 10   805   16 15 14 13 12 11 10 09 08

Harcourt
SCHOOL PUBLISHERS

Visit *The Learning Site!*
www.harcourtschool.com

How Do Plants Meet Their Needs?

Some plants have tubes that carry water, nutrients, and sugar to plant cells. The tubes transport materials in the plant. These kinds of plants are called **vascular plants.** Trees are vascular plants.

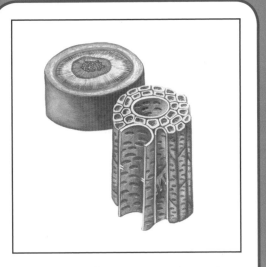

One kind of transport tissue is xylem. **Xylem** carries water and nutrients from the roots to the leaves.

Another type of transport tissue is phloem. **Phloem** carries sugar from the leaves to each of the plant's cells.

Some plants don't have vascular tissues. They are called **nonvascular plants**.

READING FOCUS SKILL

MAIN IDEA AND DETAILS

The main idea is what the text is mostly about. Details are pieces of information about the main idea.

Look for information about how plants meet their needs and details about the ways different types of plants meet their needs.

How Plants Meet Their Needs

Do you know that you need plants in order to survive? Plants give off the oxygen that is in the air you breathe. Plants also make some of the food you eat.

Plants grow in many places on Earth. They can grow in dry deserts, wet marshes, and even in the Arctic Circle. Plants can live in all of these places because they are *adapted* to live in these areas. Adapted means that they have changed over time to help them meet their needs where they live.

Seeds need sunlight, air, and water to begin growing. A seed can take in water. It gets larger and then splits open. This frees the seedling inside. Look at the picture below. You can see a small root growing downward. You can also see a small shoot growing upward. Branches may grow from buds along the sides of the stem. Branches produce leaves and more buds. From these buds, more branches grow.

◄ The seedling grows where plant cells divide quickly. Cells often divide quickly at the tip of the root and at the tip of the shoot.

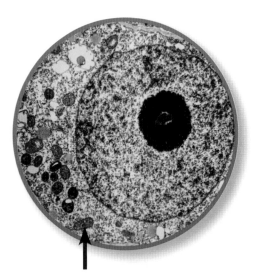

◀ You can see the parts of this cell that contain chlorophyll. It captures the sun's energy.

The new plant needs five things to stay healthy. It needs water, light, carbon dioxide, oxygen, and nutrients. The plant uses these materials to carry out life processes.

Plants are the only living things that can make their own food. They do this through a process called *photosynthesis.* For photosynthesis, plants need water, carbon dioxide, and sunlight. The roots take water from the soil. Tiny openings on the undersides of leaves take in carbon dioxide. The leaves also take in sunlight. A green chemical called *chlorophyll* captures the sun's energy. The plant uses this energy to change carbon dioxide and water to sugar and oxygen. The plant stores the sugar. It gives off oxygen into the air.

The sugar has stored chemical energy. Plants use oxygen to release this energy. This process is called *respiration.* Oxygen joins with sugar to produce energy, carbon dioxide, and water. The cells use this energy to carry out their life processes.

 **What do plants need for photosynthesis?**

Vascular Plants

All plants have the same basic needs: water, nutrients, and food. The plant takes water and nutrients in through the roots. The water and nutrients must be carried through the entire plant. The leaves make food (sugar). The food must be carried to each cell. In some plants, transport tissues carry these materials. Plants that have these tissues are called **vascular plants**.

Vascular plants can grow very large because of these transport tissues. They form a system of tubes throughout the whole plant. These tubes are made of two kinds of tissue. **Xylem** carries water and nutrients from the roots to the leaves. **Phloem** carries sugar from the leaves to each of the plant's cells.

In a vascular plant, water and minerals are pulled upward through xylem through the plant to the leaves. Sugar is moved from the leaves to all parts of the plant through phloem.

How do water and minerals move through xylem?

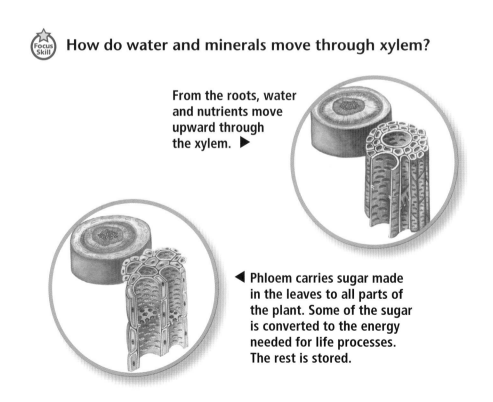

From the roots, water and nutrients move upward through the xylem. ▶

◀ Phloem carries sugar made in the leaves to all parts of the plant. Some of the sugar is converted to the energy needed for life processes. The rest is stored.

Nonvascular Plants

Plants that do not have transport tissues are **nonvascular plants**. Some examples of these plants are mosses, liverworts, and hornworts. Nonvascular plants have parts that look like leaves, stems, and roots but they are simpler than those found in vascular plants.

Nonvascular plants do not grow very large because they do not have transport tissues. These plants do not have true roots to take in water. They do not have true leaves to take in air. Instead, each cell takes in water and air.

Many nonvascular plants live in wet, shady areas like forests and swamps. These plants have leaflike parts that contain chlorophyll and produce food for the plant.

 How do nonvascular plants get materials to live?

The moss plant is a nonvascular plant. ▶

Review

 Complete this main idea statement.

1. Plants that have specialized tissues to move materials are called _____ plants.

Complete these detail statements.

2. _____ tissue moves water and minerals from a plant's roots to its leaves.

3. _____ tissue moves food from the leaves to the rest of the plant.

4. Plants make their own food through the process of _____.

How Do Plants Respond to Their Environment?

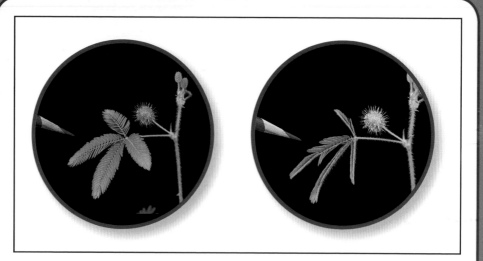

A plant's response toward or away from anything in its environment is called a **tropism**. The picture on the left shows the leaves and flower of a mimosa tree. The picture on the right shows someone touching the leaves. Notice how the leaves fold together. This is their response.

This plant is growing toward the light. It is an example of **phototropism**.

The response of plants to gravity is called **gravitropism**. Plant roots grow down into the soil and shoots grow upward, away from the pull of gravity.

READING FOCUS SKILL
CAUSE AND EFFECT

A cause is what makes something happen. An effect is what happens.

Look for causes of tropisms and rhythms in plants and the effects the tropisms and rhythms have on plants.

How Plants Respond to the Environment

Do you know that some plants move during the day? A sunflower will turn its "face" to face the sun. It will keep turning to follow the sunlight. This type of movement is called solar tracking. A **tropism** is a response of a plant toward or away from anything in its environment. The word comes from a word that means "to turn."

The response of plants to light is called **phototropism**. The tip of a plant shoot senses light. The tip then gives off chemicals. These chemicals make cells on the "dark" side of the plant grow faster. This causes the plant to bend toward the light. This is called *positive phototropism.*

Gravitropism is the response of a plant to gravity. The roots of most plants grow down, in the direction of the pull of gravity. This is called *positive gravitropism.* The shoots of seedlings grow up, away from the pull of gravity. This is called *negative gravitropism.* Gravitropism guides a plant to grow up through the soil so it can reach sunlight.

Look at the stem of the sunflower. It bends toward the sun. ▶

Can you tell why the mimosa is
called the "sensitive plant?"

Some plants respond to touch. Peas, pumpkins, and morning glories have tendrils. Tendrils look like threads. These tendrils grow toward the side of the plant that is touched. They help support the plant by wrapping themselves around objects that touch them.

Roots also respond to touch. They grow away from the things that touch them. This is a negative response to touch.

In the Arctic region, some plants have flowers that move with the sun. This is called solar tracking. Facing the sun increases the amount of energy the plant can take in. It also helps keep the flowers warm. The warm flowers attract insects. The insects pollinate the plant.

 What types of tropisms does a seedling go through as it breaks out from the seed?

Plant Rhythms

Plants have to adapt to the changing seasons. Most plants grow when it is warm. They become *dormant,* or inactive, when it is cold. Plants do this by responding to changes in temperature. Many plants also respond to changes in the length of the day.

Long-day plants make flowers when there are more than a certain number of daylight hours. It is really the amount of darkness that determines when the plants bloom. Long-day plants are really short-night plants. Sunflowers, snapdragons, and begonias are long-day plants. They bloom in late spring and early summer. This is when nights are shortest.

Short-day plants flower when there are fewer than a certain number of daylight hours. They need long periods of darkness to bloom. Short-day plants are really long-night plants. Poinsettias, strawberries, and ragweed are short-day plants.

◀ This plant is a short-day plant. It needs long periods of darkness to bloom.

Annual plants sprout, make flowers, make seeds, and die in one growing season. Impatiens, begonias, and marigolds are all annual plants.

Perennial plants return year after year. Florida paintbrush, honeysuckle, and lavender are perennial plants.

What is the effect of long periods of darkness on a long-day plant?

◄ This plant is an annual plant. It lives for only one season.

Review

Focus Skill

Complete these cause and effect statements.

1. _____ occurs when a plant turns toward a source of light.

2. Negative _____ is shown when a shoot of a seedling grows up through the soil.

3. _____-_____ plants will bloom when they have long periods of darkness.

4. A(n) _____ plant will germinate, flower, make seeds, and die within one growing season.

What Are Some Types of Plants?

A **moss** is a very small plant. It does not have vascular tissues.

In **asexual reproduction**, a new plant is formed without an egg cell and a sperm cell joining together.

Ferns are vascular plants. They reproduce without seeds. They reproduce using **spores**.

An **angiosperm** is a flowering vascular plant. Its seeds are surrounded by a fruit.

A **gymnosperm** is a plant that makes a seed that is not surrounded by a fruit. One kind of gymnosperm is a **conifer**. A conifer is a cone-bearing gymnosperm. Most trees with needle-like leaves are conifers.

READING FOCUS SKILL

COMPARE AND CONTRAST

You compare and contrast when you look for ways things are similar and different. Compare means to find the way things are similar. Contrast is to find the way things are different.

Compare and contrast the different types of plants.

Mosses, Liverworts, and Hornworts

Have you ever noticed the "furry" green growth on rocks or on the side of a tree? These are mosses. A **moss** is a very small plant. Mosses grow best in wet, shady places. You can often find them on a forest floor or near a stream or pond. Mosses do not have vascular tissues. They have structures that look like simple roots, stems, and leaves. But these are not true roots, stems, or leaves.

The rootlike structures of mosses hold the plant in the soil. But they do not take in a lot of water or nutrients. The stemlike structures of mosses have cells for carrying water and food. But they do not have tubes. The leaflike structures on mosses are green and they carry out photosynthesis. But they are only one or two cells thick and they do not have veins.

Club Moss   **Liverwort**

Mosses have a life cycle with two stages, or generations. In the *sexual reproduction* stage, egg and sperm cells join to form a new moss plant.

In the **asexual reproduction** stage, a new plant is formed without an egg cell and sperm cell joining. Mosses have a stalk with a capsule on top. This capsule makes structures that contain cells. These cells can grow into a new plant.

Liverworts and hornworts are also nonvascular plants. These plants live in the same areas as mosses. They also have life cycles with two different generations.

But liverworts and hornworts are different from mosses in some ways. They are both smaller than mosses. Liverworts do not have leaflike structures around the stem, as mosses do. They have flat, scaly leaves. Hornworts have spore-producing structures that look like horns. They have no stemlike structure or leaflike parts.

(Focus Skill) **Compare the structures of mosses with the structures of liverworts and hornworts.**

◀ Moss

Hornwort ▶

Ferns

Ferns are vascular plants. Ferns were the first plants with roots. They appeared more than 350 million years ago. About 100 million years later, Earth had more ferns than any other plant. The climate during this time was warm and humid. The ferns grew very large. As the climate cooled, there were fewer ferns. Today there are about 12,000 species of ferns. Many of these live in tropical rain forests. Light is dim on the forest floor. But ferns have wide fronds to gather enough sunlight for them to make food. Some ferns also live in the cooler, drier temperate forests.

The staghorn fern is an unusual fern. It does not have the usual lacy fronds. Its lower frond acts like a suction cup to hold the fern to a tree's bark. The rest of the fronds look like the horns of a male deer. The fern grows high in the branches of a rainforest tree. Because it is high in the tree, it gets the light it needs for photosynthesis. Its roots grow into the tree's bark. The fern does not harm the tree.

The Boston fern is a popular houseplant. ▼

Life cycle of a fern

2 This heart-shaped plant is the sexual generation of ferns. A fiddlehead is growing from it.

1

3

4

▲ Under the heart-shaped plant parts are small structures that produce egg cells and sperm cells during the sexual generation of the ferns' life cycle.

The asexual generation reproduces by spores that grow on the undersides of the ferns' leaflets.

Ferns reproduce without seeds. The life cycle of ferns, like the life cycle of mosses, has two different generations.

Look at the pictures above. Pictures **1** and **2** show the sexual generation. In both pictures, you can see structures under the heart-shaped plant part. These structures produce egg and sperm cells. After a sperm cell joins with an egg cell, a new fern plant begins to develop. This is the beginning of the asexual generation.

Pictures **3** and **4** show the asexual generation. Picture **3** shows the fern fronds. Picture **4** shows what is under the frond's leaflets. The light-colored structures are sori. Sori make spores. **Spores** contain cells that can produce a new fern. When the spores in the sori are mature, they are released and fall to the ground.

They produce the heart-shaped plant **1**. The life cycle continues.

(Focus Skill) **Contrast the structures found on a fern with the structures found on mosses.**

Gymnosperms

Most plants on Earth reproduce with seeds. A seed holds a tiny plant, called an *embryo,* and food for the embryo. A tough outer coat surrounds the seed. Two groups of plants make seeds. One group is called the gymnosperms. In a **gymnosperm**, the seeds are not surrounded by a fruit, and they don't produce flowers.

You are probably familiar with one kind of gymnosperm. Cone-bearing trees with needle-like leaves that stay green all year are gymnosperms. These kinds of trees are called **conifers**. Conifers include pine trees, juniper trees, cedar trees, and cypress trees.

Seeds of conifers develop inside cones. Conifers have two types of cones. Both types grow on the same tree. The small male cones make pollen. The female cones produce egg cells.

The egg cells mature in the female cone. When they are ready, the cone makes a sticky liquid. This liquid traps the grains of pollen, which are carried by the wind. The pollen then joins with the egg cells.

The fertilized eggs form tiny plants inside seeds. When the seeds are mature, the female cone opens up and releases them. These seeds often have wings. They can "fly" through the air like tiny helicopters. When things are right, the seeds sprout and new plants begin to grow.

 Compare and contrast pollen and egg cells.

The Welwitschia ▶
(wel•WIT•chee•uh) grows in the deserts of southwestern Africa. This plant produces only two leaves, which split as they grow, making the plant appear to have many more.

Angiosperms

Most plants are angiosperms. An **angiosperm** is a flowering vascular plant. The flowers are the plant's reproductive organs. All angiosperms reproduce sexually. Their seeds are surrounded by a fruit. Fruits form from their flowers.

Angiosperms are divided into two groups based on structures found in their seeds. These structures are seed leaves. Seed leaves store food for the embryo. Seeds can have one seed leaf or two seed leaves.

 Compare and contrast gymnosperms and angiosperms.

Peonies are angiosperms. ▶

Review

Complete these compare and contrast statements.

1. Mosses have _____ that move food. Vascular plants have _____ that transport food.

2. Liverworts and hornworts are both _____ plants. They both live in the same type of _____. They both have life cycles with two different _____.

3. _____ are plants whose seeds are surrounded by a fruit. _____ are plants whose seeds are not surrounded by a fruit.

How Do Angiosperms Reproduce?

VOCABULARY

pollination
fruit

Pollination is the first step in angiosperm reproduction. It is the transfer of pollen.

A **fruit** is the ripened ovary of a
flowering plant.

Flower Parts

   Angiosperms have flowers. Flowers have many different parts. The largest parts of the flower are its *petals.* They are often brightly colored. The petals help protect other parts of the flower. Below the petals are the *sepals.* They often look like little green leaves. Sepals cover the bud. They protect it as it develops. After the flower blooms, the sepals spread apart.

In the center of the flower are several stemlike structures called *stamens*. Stamens are the flower's male reproductive organs. A stamen has two parts. The *anther* produces pollen grains. The *filament* connects the anther to the plant.

The *pistil* is in the center of the flower. The pistil is the flower's female reproductive organ. It has three parts. At the top is the *stigma.* This is a sticky structure. Pollen grains fall on it. In the middle is the *style.* It connects the stigma to the *ovary.* Inside the ovary are *ovules* that contain egg cells. The egg cells will develop into seeds.

 **What are the three parts of the pistil?**

Look for the reproductive organs in the daffodil plant. ▶

Flowers and Pollination

Have you ever noticed bees buzzing around flowers? They are working. They are helping plants reproduce. They are transferring pollen from one plant to another.

Pollination is the transfer of pollen. It is the first step in angiosperm reproduction. It happens when pollen from an anther lands on the stigma of a flower of the same kind. *Self-pollination* happens when pollen is transferred within the same flower. *Cross-pollination* happens when pollen is transferred from the anther of one flower to the stigma of another flower.

Butterflies, bees, hummingbirds, and other animals help transfer pollen to the stigmas of flowers.

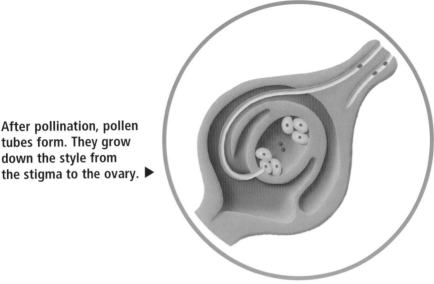

After pollination, pollen tubes form. They grow down the style from the stigma to the ovary. ▶

After pollination, a tube grows from each pollen grain. You can see this in the picture above. The tube goes down the style to the ovary. Then the sperm cells from the pollen grains go down the tube to join with the egg cells. This is *fertilization.*

The fertilized egg cells then develop into seeds. As this happens, the ovary forms a **fruit** that surrounds the seeds.

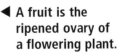

◀ A fruit is the ripened ovary of a flowering plant.

Plants have adaptations to improve their chances of pollination. Some flowers have brightly colored petals and strong scents to attract animals. They produce a liquid called *nectar* that some animals use as food.

Plants that are pollinated by animals have flowers that make heavy, sticky pollen. The pollen gets caught on a passing animal's body. Then the animal rubs up against the pistil of another flower. The pollen rubs off onto the stigma. Insects, birds, bats, and other animals can move pollen from one flower to another.

Plants that are not pollinated by animals depend on the wind to carry pollen from flower to flower. These plants make a lot of light pollen. The wind carries this pollen from the anther of one flower to the stigma of another flower of the same kind. Oak trees, maple trees, corn, and other grasses are pollinated by the wind. The flowers of these plants are small. They do not have a scent because they do not need to attract animals.

Seeds are more likely to grow well away from the parent plant. Some seeds are spread by wind. Some are spread by animals.

(Focus Skill) What happens after pollination occurs?

This dog can carry seeds far away before they are rubbed or scratched off. ▼

The seeds of the dandelion are light enough to be carried by the wind. ▼

Fruits and Seeds

Angiosperms are divided into two groups based on their seed structure. Seeds have seed leaves inside them. Seed leaves store food for the embryo. Monocots have one seed leaf. Dicots have two seed leaves. Look at the table below to see some of the other ways monocots and dicots are different from one another.

How are monocots and dicots different?

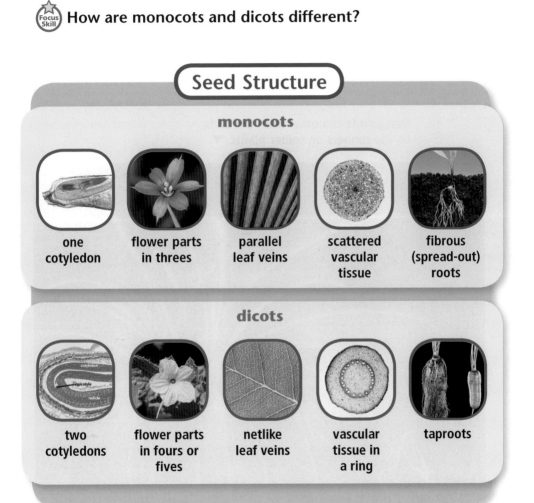

Seed Structure

monocots

| one cotyledon | flower parts in threes | parallel leaf veins | scattered vascular tissue | fibrous (spread-out) roots |

dicots

| two cotyledons | flower parts in fours or fives | netlike leaf veins | vascular tissue in a ring | taproots |

Asexual Reproduction

Some plants reproduce without using seeds. This is a form of asexual reproduction. Asexual reproduction makes plants that are identical to the parent plant.

You can grow some plants from leaf cuttings. Place the leaves from the parent plant in water or soil. The leaves may then grow into new plants.

Tubers are plants with underground stems that swell and store food. Their "eyes" can grow into new plants. Potatoes are tubers.

New plants can grow from the buds on the runners on spider plants. ▼

You can also use *grafting* to grow some plants, like roses. You join the parts of two plants. This gives you a plant that has the characteristics of both parents.

 **Describe the process of grafting.**

 **Parts of two plants have been joined to make one plant in the process known as *grafting*.** ▶

Review

Focus Skill

Complete this main idea statement.

1. Flowers have bright petals, nectar, and sticky pollen grains to help them _____ .

Complete these detail statements.

2. A _____ _____ grows from the stigma to the ovary.

3. Sperm cells join with egg cells in the ovules during _____.

4. Spider plants reproduce using _____. This is a form of asexual reproduction.

GLOSSARY

angiosperm (AN•jee•oh•sperm) A flowering vascular plant whose seeds are surrounded by a fruit.

asexual reproduction (ay•SEK•shoo•uhl ree•pruh•DUHK•shuhn) Type of reproduction in which a new organism is formed without the joining of a sperm cell and an egg cell.

conifer (KAHN•uh•fer) A type of gymnosperm whose seeds develop inside a cone.

fern (FERN) A vascular plant that reproduces without seeds.

fruit (FROOT) The ripened ovary of a flowering plant.

gravitropism (gra•VIH•truh•piz•uhm) The growth response of plants to gravity.

gymnosperm (JIM•noh•sperm) A vascular plant that produces seeds that are not surrounded by fruit.

moss (MAHS) A small plant that does not have vascular tissues or true roots, stems, or leaves.

nonvascular plant (nahn•VAS•kyuh•ler PLANT) A plant that lacks tissues for carrying water, food, and nutrients.

phloem (FLOH•em) Plant tissue that carries food (sugar) from the leaves to each of the plant's cells.

phototropism (foh•TAH•truh•piz•uhm) The growth response of plants to light.

pollination (pahl•uh•NAY•shuhn) The first step of angiosperm reproduction, during which pollen from an anther lands on a stigma of a flower of the same kind.

spore (SPAWR) A structure containing cells that can grow into a new plant without joining with other cells.

tropism (TROH•piz•uhm) A response of a plant toward or away from something in its environment.

vascular plant (VAS•kyuh•ler PLANT) A plant that has transport tissues for carrying water, food, and nutrients to its cells.

xylem (ZY•luhm) Plant tissue that carries water and nutrients from the roots to the leaves.